1996 No. 192

HEALTH AND SAFETY

The Equipment and Protective Systems Intended for Use in Potentially Explosive Atmospheres Regulations 1996

Made - - - -	*1st February 1996*
Laid before Parliament	*5th February 1996*
Coming into force	*1st March 1996*

ARRANGEMENT OF REGULATIONS

PART I

PRELIMINARY

PART II

APPLICATION

PART III

GENERAL REQUIREMENTS

PART IV

ENFORCEMENT

SCHEDULES

The Secretary of State, being a Minister designated(**a**) for the purposes of section 2(2) of the European Communities Act 1972(**b**) in relation to measures relating to equipment and protective systems intended for use in potentially explosive atmospheres, in exercise of the powers conferred on him by that section and of all his other enabling powers, hereby makes the following Regulations:

PART I

PRELIMINARY

Citation, commencement, extent and revocation

1.—(1) These Regulations, which extend to Great Britain, may be cited as the Equipment and Protective Systems Intended for Use in Potentially Explosive Atmospheres Regulations 1996 and shall come into force on 1st March 1996.

(2) The Regulations specified in the first column of Schedule 1 hereto are hereby revoked with effect from the date specified in the second column of that Schedule.

Interpretation

2.—(1) In these Regulations–

(a) the "ATEX Directive" means Directive 94/9/EC of the European Parliament and the Council on the approximation of the laws of the Member States concerning equipment and protective systems intended for use in potentially explosive atmospheres(**c**); and

(b) except for the references to the European Communities in the definition of "the Commission" and in relation to the Official Journal, a reference to the Community includes a reference to the EEA, and a reference to a member State includes a reference to an EEA State: for the purposes of this sub-paragraph–

   (i) the "EEA" means the European Economic Area;

   (ii) an "EEA State" means a State which is a Contracting Party to the EEA Agreement(**d**); and

   (iii) the "EEA Agreement" means the Agreement on the European Economic Area signed at Oporto on 2nd May 1992 as adjusted by the Protocol signed at Brussels on 17th March 1993.(**e**)

(2) In these Regulations, unless the context otherwise requires–

"CE marking" or "CE conformity marking" is a reference to a marking consisting of the initials "CE" in the form shown in Schedule 2 hereto;

"the Commission" means the Commission of the European Communities;

"component" has the meaning given by regulation 3(2)(d) below;

"devices" and "devices referred to in Article 1(2)" have the meaning given by regulation 3(2)(c) below;

"enforcement authority" means the Health and Safety Executive established under section 10 of the Health and Safety at Work etc. Act 1974(**f**);

"essential health and safety requirements" means the requirements in Annex II of the ATEX Directive which is set out in Schedule 3 hereto;

"equipment" has the meaning given by regulation 3(2)(a) below;

(**a**) S.I. 1995/751.
(**b**) 1972 c.68.
(**c**) OJ No. L100, 19.4.94, p.1.
(**d**) The EEA Agreement came into force in relation to Liechtenstein on 1st May 1995 (OJ No. L86, 20.4.95, p. 58).
(**e**) The application of the ATEX Directive was extended to the EEA from 1st December 1994 by virtue of Decision 14/94 of the EEA Joint Committee (OJ No. L325, 17.12.94, p. 65) which inserted a reference to that Directive at point 7A in Chapter X of Annex II to the EEA Agreement.
(**f**) 1974 c.37.

"equipment-category" and "category" in relation to an equipment-group shall be construed by reference to the criteria set out in Annex I of the ATEX Directive which is set out in Schedule 4 hereto;

"equipment group I" means equipment intended for use in underground parts of mines, and to those parts of surface installations of such mines, liable to be endangered by firedamp and/or combustible dust;

"equipment group II" means equipment intended for use in places, other than those which are specified for equipment group I, liable to be endangered by explosive atmospheres;

"explosive atmosphere" and "potentially explosive atmosphere" shall be construed respectively in accordance with regulation 3(2)(f) and (g) below;

"harmonized standard" means a technical specification adopted by the European Committee for Standardisation or the European Committee for Electrotechnical Standardisation or both, upon a mandate from the Commission in accordance with Council Directive 83/189/EEC of 28th March 1983 laying down a procedure for the provision of information in the field of technical standards and regulations(a), and of which the reference number is published in the Official Journal of the European Communities;

"intended use" shall be construed in accordance with regulation 3(2)(e) below;

"notified body" shall be construed in accordance with regulation 11 below;

"protective systems" has the meaning given by regulation 3(2)(b) below;

"relevant essential health and safety requirements" in relation to equipment, a protective system or device means those provisions of the essential health and safety requirements which are applicable to that particular equipment, protective system or device, account being taken of its intended use;

"responsible person" means, in relation to equipment, a protective system, device or component,

    (a) the manufacturer of that equipment, protective system, device or component;

    (b) the manufacturer's authorized representative established in the Community; or

    (c) where the manufacturer is not established in the Community and either–

        (i) he has not appointed an authorized representative established in the Community; or

        (ii) his authorized representative established in the Community is not the person who places that equipment, protective system, device or component on the market,

    the person who places it on the market in the Community;

"safe" in relation to equipment, a protective system or a device means that, when properly installed and maintained and used for its intended purpose, it does not endanger the health and safety of persons and, where appropriate, domestic animals or property and cognate expressions shall be construed accordingly;

"standard" or "standard referred to in Article 5" means a technical specification approved by a recognised standardising body for repeated or continuous application, with which compliance is not compulsory: and, for the avoidance of doubt, this definition includes a harmonized standard or a transposed harmonized standard;

"supply" includes offering to supply, agreeing to supply, exposing for supply and possessing for supply and cognate expressions shall be construed accordingly; and

"transposed harmonized standard" means a national standard of a member State which transposes a harmonized standard.

(a) OJ No. L109, 26.4.83, p. 8. Council Directive 83/189/EEC was amended by Council Directive 88/182/EEC (OJ No. L81, 26.3.88, p. 75), Commission Decision 92/400/EEC (OJ No. L221, 6.8.92, p. 55) and Directive 94/10/EC of the European Parliament and the Council (OJ No. L100, 19.4.94, p. 30).

PART II

APPLICATION

Equipment, Protective Systems, Devices and Components

3.—(1) Subject to regulations 4 and 5 below, these Regulations apply to equipment and protective systems intended for use in potentially explosive atmospheres, devices and components.

(2) For the purposes of these Regulations,

  (a) "equipment" means machines, apparatus, fixed or mobile devices, control components and instrumentation thereof and detection or prevention systems which, separately or jointly, are intended for the generation, transfer, storage, measurement, control and conversion of energy or the processing of material and which are capable of causing an explosion through their own potential sources of ignition;

  (b) "protective systems" means design units which are intended to halt incipient explosions immediately and/or to limit the effective range of explosion flames and explosion pressures; protective systems may be integrated into equipment or separately placed on the market for use as autonomous systems;

  (c) "devices" means safety devices, controlling devices and regulating devices intended for use outside potentially explosive atmospheres but required for or contributing to the safe functioning of equipment and protective systems with respect to the risks of explosion;

  (d) "component" means any item essential to the safe functioning of equipment and protective systems but with no autonomous function;

  (e) "intended use" means the use of equipment, protective systems, and devices in accordance with the equipment group and category and with all the information supplied by the manufacturer which is required for the safe functioning of equipment, protective systems and devices;

  (f) "explosive atmosphere" means the mixture with air, under atmospheric conditions, of flammable substances in the form of gases, vapours, mists or dusts in which, after ignition has occurred, combustion spreads to the entire unburned mixture; and

  (g) "potentially explosive atmosphere" means an atmosphere which could become explosive due to local and operational conditions.

Excluded equipment, protective systems, devices and components

4. These Regulations do not apply–

  (a) to the equipment, protective systems and devices specified in Schedule 5 hereto; and

  (b) components for the equipment, protective systems or devices referred to in paragraph (a) above.

Exclusion until 30th June 2003 of equipment and protective systems complying with health and safety provisions in force on 23rd March 1994

5.—(1) Subject to paragraph (2) below, these Regulations do not apply to equipment or a protective system placed on the market in the Community on or before 30th June 2003 which complies with any health and safety provisions with which it would have been required to comply for it to be lawfully placed on the market in Great Britain on 23rd March 1994.

(2) The exclusion provided in paragraph (1) above does not apply in the case of equipment or a protective system which–

  (a) unless required to bear the CE marking pursuant to any other Community obligation, bears the CE marking or an inscription liable to be confused therewith; or

  (b) bears or is accompanied by any other indication, howsoever expressed, that it complies with the ATEX Directive.

(3) In this regulation, "health and safety provisions" means any requirement imposed by an enactment which has the same, or substantially the same, effect as any of the essential health and safety requirements and which would, but for the provisions of this regulation, be applicable to that equipment or protective system for the purposes of complying with these Regulations.

PART III

GENERAL REQUIREMENTS

General duty relating to the placing on the market of equipment, protective systems or devices by a responsible person

6.—(1) Subject to regulation 9 below, no person who is a responsible person shall place on the market any equipment, protective system or device to which these Regulations apply unless the requirements of paragraph (2) below have been complied with in relation thereto.

(2) The requirements of this paragraph, in respect of any equipment, protective system or device, are that–

    (a) it satisfies the relevant essential health and safety requirements and, for the purpose of satisfying those requirements,

        (i) where a transposed harmonized standard covers one or more of the essential health and safety requirements, any equipment, protective system or device constructed in accordance with that transposed harmonized standard shall be presumed to comply with that or, as the case may be, those essential health and safety requirements; and

        (ii) a certificate of conformity to the harmonized standards specified in the Electrical Equipment for Explosive Atmospheres (Certification) Regulations 1990(a) ("the 1990 Regulations") and obtained in accordance with the procedures for obtaining such certificates in the 1990 Regulations shall continue to be valid for the purposes of these Regulations until 30th June 2003 (unless it expires before that date) in respect of any electrical equipment, as defined in the 1990 Regulations, which conforms to the type covered by the said certificate;

    (b) the appropriate conformity assessment procedure, in accordance with regulation 10(1) below, has been carried out–

        (i) by the manufacturer; or

        (ii) where permitted by that procedure, wholly or partly as the case may be, by the manufacturer's authorized representative established in the Community,

    save that,

        (aa) where the procedure in Annex III, VI or VIII of the ATEX Directive (which are respectively set out in Schedules 6, 9 and 11 hereto) is part of or, as the case may be, is the appropriate conformity assessment procedure; and

        (bb) the person placing the equipment, protective system or device on the market is neither the manufacturer nor his authorized representative established in the Community,

    the obligation to retain the technical documentation, required as part of that appropriate conformity assessment procedure, shall be fulfilled by the person who places that equipment, protective system or device on the market;

    (c) the CE marking has been affixed to it by the manufacturer or his authorized representative established in the Community in accordance with Schedule 2 hereto and Schedule 2 shall have effect for that purpose; and

    (d) it is in fact safe.

(a) S.I. 1990/13; relevant amending instruments are S.I. 1990/2377, 1991/2826, 1995/1186.

General duty relating to the supply of equipment, protective systems or devices by a person other than the responsible person

7.—(1) Subject to paragraph (2) below, it shall be the duty of any person who supplies any equipment, protective system or device to which these Regulations apply, but who is not a person to whom regulation 6 above applies, to ensure that that equipment, protective system or device is safe.

(2) Without prejudice to any other safety requirement which may apply in respect of such equipment, protective system or device, this regulation does not apply to–

(a) equipment, a protective system or device which has been placed on the market in the Community before 1st March 1996; or

(b) the supply of any equipment, protective system or device which has previously been put into service in the Community.

General duty relating to the placing on the market of components by a responsible person

8.—(1) Subject to regulation 9(a) below, no person who is a responsible person shall place on the market any component to which these Regulations apply unless the requirements of paragraph (2) below have been complied with in relation thereto.

(2) The requirements of this paragraph, in respect of any component, are that–

(a) the appropriate conformity assessment procedure, in accordance with regulation 10(2) below, has been carried out by the person specified in regulation 6(2)(b) above; and

(b) it is accompanied by a certificate which has been issued by the manufacturer or his authorized representative established in the Community and which–

(i) declares the conformity of the component with the provisions of the ATEX Directive which apply to it; and

(ii) states its characteristics and how it must be incorporated into equipment or protective systems to assist compliance with the essential requirements applicable to finished equipment or protective systems.

Exceptions to placing on the market in respect of certain equipment, protective systems, devices and components

9. For the purposes of regulation 6 or 8 above, equipment, a protective system, device or, in the case of paragraph (a) below, a component shall not be regarded as being placed on the market–

(a) where that equipment, protective system, device or component

(i) will be put into service in a country outside the Community; or

(ii) is imported into the Community for re-export to a country outside the Community,

save that this paragraph shall not apply if the CE marking, or any inscription liable to be confused therewith, is affixed thereto; or

(b) by the exhibition at trade fairs and exhibitions of that equipment, protective system or device, provided that where the provisions of these Regulations are not satisfied–

(i) a notice is displayed in relation to the equipment, protective system or device in question to the effect–

(aa) that it does not satisfy those provisions; and

(bb) that it may not lawfully be placed on the market until the responsible person has ensured that those provisions are satisfied; and

(ii) adequate safety measures are taken to ensure the safety of persons.

Conformity assessment procedures

10.—(1) Subject to paragraphs (4) and (5) below, for the purposes of regulation 6(2)(b) above, the appropriate conformity assessment procedure shall–

(a) in the case of equipment and, where necessary, a device, be determined in accordance with paragraph (3) below by reference to the equipment-group and equipment-category of that particular equipment or, as the case may be, device; and

(b) in the case of an autonomous protective system, be the procedure set out in paragraph (3)(a) or (d) below.

(2) For the purposes of regulation 8(2)(a) above, in the case of a component the appropriate conformity assessment procedure shall be the procedure set out in paragraph (3) below, which relates to the equipment or protective system into which that component is to be incorporated, with the exception of the affixing of the CE marking.

(3) The procedures referred to in paragraphs (1) and (2) above are as follows:

(a) without prejudice to sub-paragraph (d) below, in the case of equipment-group I and II, equipment-category M 1 and 1, the manufacturer or his authorized representative established in the Community must, in order to affix the CE marking, follow the EC type-examination procedure (referred to in Annex III of the ATEX Directive, which is set out in Schedule 6 hereto), in conjunction with:

(i) the procedure relating to production quality assurance (referred to in Annex IV of the ATEX Directive, which is set out in Schedule 7 hereto); or

(ii) the procedure relating to product verification (referred to in Annex V of the ATEX Directive, which is set out in Schedule 8 hereto);

(b) without prejudice to sub-paragraph (d) below, in the case of equipment-group I and II, equipment-category M 2 and 2,

(i) in the case of internal combustion engines and electrical equipment in these groups and categories, the manufacturer or his authorized representative established in the Community shall, in order to affix the CE marking, follow the EC-type examination procedure (referred to in Annex III of the ATEX Directive, which is set out in Schedule 6 hereto), in conjunction with:

(aa) the procedure relating to conformity to type referred to in Annex VI of the ATEX Directive (which is set out in Schedule 9 hereto); or

(bb) the procedure relating to product quality assurance referred to in Annex VII of the ATEX Directive (which is set out in Schedule 10 hereto); and

(ii) in the case of other equipment in these groups and categories, the manufacturer or his authorized representative established in the Community must, in order to affix the CE mark, follow the procedure relating to internal control of production (referred to in Annex VIII of the ATEX Directive, which is set out in Schedule 11 hereto) and communicate the dossier, provided for in paragraph 3 of Annex VIII, to a notified body, which shall acknowledge receipt of it as soon as possible and shall retain it;

(c) without prejudice to sub-paragraph (d) below, in the case of equipment-group II, equipment-category 3, the manufacturer or his authorized representative established in the Community must, in order to affix the CE marking, follow the procedure relating to internal control of production referred to in Annex VIII of the ATEX Directive (which is set out in Schedule 11 hereto); or

(d) in the case of equipment-groups I and II, as an alternative to the procedures referred to in sub-paragraphs (a), (b) and (c) above, the manufacturer or his authorized representative established in the Community may, in order to affix the CE marking, follow the procedure relating to CE unit verification (referred to in Annex IX of the ATEX Directive, which is set out in Schedule 12 hereto).

(4) In the case of equipment, protective systems or devices, the manufacturer or his authorized representative established in the Community may, in order to affix the CE marking, follow the procedure relating to internal control of production (referred to in Annex VIII of the ATEX Directive, which is set out in Schedule 11 hereto) with regard to the safety aspects referred to in point 1.2.7 of Annex II of the ATEX Directive (which is set out in Schedule 3 hereto).

(5) Notwithstanding the previous paragraphs of this regulation, the Secretary of State may, on a duly justified request, authorize the placing on the market and putting into service of equipment, protective systems and individual devices refered to in Article 1(2) in respect of which the procedures referred to in the previous paragraphs have not been applied and the use of which is in the interests of protection.

(6) Documents and correspondence relating to the procedures referred to in the above-mentioned paragraphs shall be drawn up in one of the official languages of the member States in which those procedures are being applied or in a language accepted by the notified body to which an application is made pursuant to one of those procedures.

Notified bodies

11. For the purposes of these Regulations, a notified body is a body which has been appointed to carry out one or more of the conformity assessment procedures specified in Article 8 of the ATEX Directive and referred to in regulation 10 above which has been–

    (a) appointed as a notified body in Great Britain pursuant to regulation 12 below;

    (b) appointed as a notified body in Northern Ireland; or

    (c) appointed by a member State other than the United Kingdom,

and in the case of (a), (b) and (c) above has been notified by the member State concerned to the Commission and the other member States pursuant to Article 9(1) of the ATEX Directive.

Notified bodies appointed by the Secretary of State

12.—(1) The Secretary of State may from time to time appoint such qualified persons as he thinks fit to be notified bodies for the purposes of these Regulations.

(2) An appointment–

    (a) may relate to all descriptions of equipment, protective systems, devices or components or such descriptions (which may be framed by reference to any circumstances whatsoever) of equipment, protective systems, devices or components as the Secretary of State may from time to time determine;

    (b) may be made subject to such conditions as the Secretary of State may from time to time determine, and such conditions may include conditions which are to apply upon or following termination of the appointment;

    (c) shall, without prejudice to the generality of sub-paragraph (b) above, require that body, subject to paragraph (4) below, to carry out the procedures and specific tasks for which it has been appointed including (where so provided as part of those procedures) surveillance to ensure that the manufacturer duly fulfils the obligations arising out of the relevant quality assurance procedure;

    (d) shall be terminated–

        (i) if it appears to the Secretary of State that the notified body is no longer a qualified person; or

        (ii) upon 90 days' notice in writing to the Secretary of State, at the request of the notified body; and

    (e) may be terminated if it appears to the Secretary of State that any of the conditions of the appointment are not complied with.

(3) Subject to paragraph (2)(d) and (e) above, an appointment under this regulation may be for the time being or for such period as may be specified in the appointment.

(4) A notified body appointed by the Secretary of State shall not be required to carry out the functions referred to in paragraph (2)(c) above if–

    (a) the documents submitted to it in relation to carrying out such functions are not in English or another language acceptable to that body;

    (b) the person making the application has not submitted with its application the amount of the fee which the body requires to be submitted with the application pursuant to regulation 13 below; or

    (c) the body reasonably believes that, having regard to the number of applications made to it in relation to its appointment under these Regulations which are outstanding, it will be unable to commence the required work within 3 months of receiving the application.

(5) If for any reason the appointment of a notified body is terminated under this regulation, the Secretary of State may authorise another notified body to take over its functions in respect of such cases as he may specify.

(6) A notified body which is responsible, as part of any of the conformity assessment procedures referred to in regulation 10 above, for the assessment of the conformity of electrical equipment placed on the market before 1st July 2003, shall take account of the results of tests and verifications already carried out in respect of the harmonized standards which are applicable under–

(a) Council Directive 76/117/EEC(**a**) and Council Directive 79/196/EEC(**b**); or

(b) Council Directive 82/130/EEC(**c**).

(7) If a notified body, to which an application has been made for an EC type-examination certificate pursuant to the EC type-examination procedure (referred to in Annex III of the ATEX Directive and set out in Schedule 6 hereto), is not satisfied that the requirements for such a certificate are met and is minded to refuse to issue an EC type-examination certificate, it shall–

(a) inform the applicant in writing of the reasons why it proposes to refuse to issue an EC type-examination certificate;

(b) give the applicant the opportunity, within a reasonable period, of making representations as to why it should not be refused; and

(c) if, after considering any representations made pursuant to sub-paragraph (b) above, it remains unsatisfied in respect of those requirements, it shall–

(i) notify its decision in writing to the applicant stating the grounds on which the refusal is based; and

(ii) inform the applicant in writing of the procedure which it has established whereby an appeal may be made against that decision.

(8) In this regulation–

"qualified person" means a person (which may include the Secretary of State) who meets the minimum criteria; and

"minimum criteria" means the criteria set out in Annex XI of the ATEX Directive (minimum criteria to be taken into account by member States for the notification of bodies)(**d**).

Fees

13.—(1) Without prejudice to the power of the Secretary of State, where he is a notified body, to charge fees pursuant to regulations made under section 56 of the Finance Act 1973(**e**) and subject to paragraph (2) below, a notified body appointed by the Secretary of State, other than the Secretary of State, may charge such fees in connection with, or incidental to, carrying out its duties in relation to the functions referred to in regulation 12(2)(c) above as it may determine; provided that such fees shall not exceed the sum of the following–

(a) the costs incurred or to be incurred by the notified body in performing the relevant function; and

(b) an amount on account of profit which is reasonable in the circumstances having regard to–

(i) the character and extent of the work done or to be done by the body on behalf of the applicant; and

(ii) the commercial rate normally charged on account of profit for that work or similar work.

(2) The power in paragraph (1) above includes the power to require the payment of fees or a reasonable estimate thereof in advance of carrying out the work requested by the applicant.

(**a**) OJ No. L24, 30.1.76, p. 45.

(**b**) OJ No. L43, 20.2.79, p. 20. Council Directive 79/196/EEC was adapted to technical progress by Commission Directives 84/47/EEC (OJ No. L31, 2.2.84, p. 19), 88/571/EEC (OJ No. L311, 17.11.88, p. 46) and 94/26/EC (OJ No. L157, 24.6.94, p. 33) and was amended by Council Directives 88/665/EEC (OJ No. L382, 31.12.88, p. 42) and 90/487/EEC (OJ No. L270, 2.10.90, p. 23).

(**c**) OJ No. L59, 2.3.82, p. 10. Council Directive 82/130/EEC was adapted to technical progress by Commission Directives 88/35/EEC (OJ No. L20, 26.1.88, p. 28), 91/269/EEC (OJ No. L134, 29.5.91, p. 51) and 94/44/EC (OJ No. L248, 23.9.94, p. 22).

(**d**) Notified bodies meeting the assessment criteria laid down in the relevant harmonized standards are presumed to meet the minimum criteria.

(**e**) 1973 c.51.

Supplementary provisions

Conditions for equipment, protective systems, devices and components being taken to comply with the provisions of the ATEX Directive

14.—(1) Subject to paragraph (2) below,

(a) equipment, a protective system or device–

   (i) which is accompanied by an EC declaration of conformity–

      (aa) issued in respect of it by the manufacturer or his authorized representative established in the Community; and

      (bb) containing the elements set out in Schedule 13 hereto; and

   (ii) to which the CE marking is affixed in accordance with regulation 6(2)(c) above; or

(b) a component which is accompanied by a certificate, which has been issued in accordance with regulation 8(2)(b) above,

shall be taken to comply with all the provisions of the ATEX Directive including the appropriate conformity assessment procedure specified in regulation 10 above, unless there are reasonable grounds for suspecting that it does not so comply.

(2) Paragraph (1) above does not apply–

(a) in relation to the enforcement authority where the responsible person fails or refuses to make available to the enforcement authority the documentation which he is required, by the conformity assessment procedure which applies to that equipment, protective system, device or component, to retain or a copy thereof; or

(b) in the case of equipment, a protective system or device

   (i) which is supplied in the circumstances described in regulation 7(2)(b) above; and

   (ii) to which the CE marking is indelibly affixed.

PART IV
ENFORCEMENT

Application of Schedule 14

15.—(1) Subject to paragraph (2) below, Schedule 14 shall have effect for the purposes of providing for the enforcement of these Regulations and for matters incidental thereto.

(2) Except in the case of equipment, a protective system or device which, in the opinion of the enforcement authority, is not safe, where the enforcement authority has reasonable grounds for suspecting that the CE marking has not been correctly affixed to equipment, a protective system or device, as the case may be, it may give notice in writing to the responsible person who placed that equipment, protective system or device, on the market and, subject to paragraph (3) below, no action pursuant to Schedule 14 may be taken, and no proceedings may be brought pursuant to regulation 16 below, in respect of that equipment, protective system or device, as the case may be, until such notice has been given and the person to whom it is given has failed to comply with its requirements.

(3) Notwithstanding the provisions of paragraph (2) above, for the purpose of ascertaining whether or not the CE marking has been correctly affixed, action may be taken pursuant to section 20 of the Health and Safety at Work etc. Act 1974(a) as it is applied by Schedule 14.

(4) Notice which is given under paragraph (2) above shall–

(a) state that the enforcement authority suspects that the CE marking has not been correctly affixed to the equipment, protective system or device, as the case may be;

(b) specify the respect in which it is so suspected and give particulars thereof;

(a) 1974 c.37.

(c) require the person to whom the notice is given–
  (i) to secure that any equipment, protective system or device, as the case may be, to which the notice relates conforms as regards the provisions concerning the correct affixation of the CE marking within such period as may be specified in the notice; or
  (ii) to provide evidence within that period, to the satisfaction of the enforcement authority, that the CE marking has been correctly affixed; and
(d) warn that person that if the non-conformity continues after (or if satisfactory evidence has not been provided within) the period specified in the notice, further action may be taken under the Regulations in respect of that equipment, protective system or device, as the case may be, or any equipment, protective system or device of the same type placed on the market by that person.

(5) For the purposes of this regulation, the CE marking is correctly affixed to equipment, a protective system or device, as the case may be, if–
(a) it has been affixed in accordance with regulation 6(2)(c) above; and
(b) the appropriate conformity assessment procedure has been carried out in respect of that equipment, protective system or device in accordance with regulation 6(2)(b) above.

Offences

16. Any person who contravenes or fails to comply with regulation 6, 7 or 8 above shall be guilty of an offence.

Penalties

17. A person guilty of an offence under regulation 16 above shall be liable on summary conviction–
(a) to imprisonment for a term not exceeding 3 months; or
(b) to a fine not exceeding level 5 on the standard scale,
or to both.

Defence of due diligence

18.—(1) Subject to the following provisions of this regulation, in proceedings against any person for an offence under regulation 16 above it shall be a defence for that person to show that he took all reasonable steps and exercised all due diligence to avoid committing the offence.

(2) Where in any proceedings against any person for such an offence the defence provided by paragraph (1) above involves an allegation that the commission of the offence was due–
(a) to the act or default of another; or
(b) to reliance on information given by another,
that person shall not, without the leave of the court, be entitled to rely on the defence unless, not less than seven clear days before the hearing of the proceedings (or, in Scotland, the trial diet), he has served a notice under paragraph (3) below on the person bringing the proceedings.

(3) A notice under this paragraph shall give such information identifying or assisting in the identification of the person who committed the act or default or gave the information as is in the possession of the person serving the notice at the time he serves it.

(4) It is hereby declared that a person shall not be entitled to rely on the defence provided by paragraph (1) above by reason of his reliance on information supplied by another, unless he shows that it was reasonable in all the circumstances for him to have relied on the information, having regard in particular–
(a) to the steps which he took, and those which might reasonably have been taken, for the purpose of verifying the information; and
(b) to whether he had any reason to disbelieve the information.

Liability of persons other than the principal offender

19.—(1) Where the commission by any person of an offence under regulation 16 above is due to the act or default committed by some other person in the course of any business of his, the other person shall be guilty of the offence and may be proceeded against and punished by virtue of this paragraph whether or not proceedings are taken against the first-mentioned person.

(2) Where a body corporate is guilty of an offence under these Regulations (including where it is so guilty by virtue of paragraph (1) above) in respect of any act or default which is shown to have been committed with the consent or connivance of, or to be attributable to any neglect on the part of, any director, manager, secretary or other similar officer of the body corporate or any person who was purporting to act in any such capacity he, as well as the body corporate, shall be guilty of that offence and shall be liable to be proceeded against and punished accordingly.

(3) Where the affairs of a body corporate are managed by its members, paragraph (2) above shall apply in relation to the acts and defaults of a member in connection with his functions of management as if he were a director of the body corporate.

(4) In this regulation, references to a "body corporate" include references to a partnership in Scotland and, in relation to such partnership, any reference to a director, manager, secretary or other similar officer of a body corporate is a reference to a partner.

Amendment and disapplication of law in Great Britain

General provisions

20.—(1) In sub-paragraph (d) of regulation 19(2) of The Electricity at Work Regulations 1989(**a**), for the words "Commission Directive 88/35/EEC(**b**)" there shall be substituted the words "Commission Directives 88/35/EEC(**c**), 91/269/EEC(**d**) and 94/44/EC(**e**)" and, for the purposes of the enforcement of that sub-paragraph, this substitution shall have effect as if it had been made under section 15 of the Health and Safety at Work etc. Act 1974(**f**).

(2) Subject to regulation 1(2) above, in respect of any equipment, protective system, device or component to which these Regulations apply, there shall be disapplied–
  (a) in the Coal and Other Mines (Locomotives) Regulations 1956(**g**)–
    (i) paragraphs (1) and (2) of regulation 3; and
    (ii) regulations 4 and 5;
  (b) in both The Bentinck Mine (Diesel Engined Stone Dusting Machine) Regulations 1976(**h**) and The Point of Ayr Mine (Diesel Vehicles) Regulations 1980(**i**), regulation 4; and
  (c) in the Regulations specified in the first column of Schedule 15 hereto, the regulations respectively specified in the third column of that Schedule.

Consequential amendment of the Provision and Use of Work Equipment Regulations 1992

21.—(1) At the end of Schedule 1 to the Provision and Use of Work Equipment Regulations 1992(**j**), there shall be added the following–
  "37. European Parliament and Council Directive 94/9/EC on the approximation of the laws of the Member States concerning equipment and protective systems intended for use in potentially explosive atmospheres (OJ No. L100, 19.4.94, p. 1).".

(**a**) S.I. 1989/635.
(**b**) OJ No. L20, 26.1.88, p. 28.
(**c**) OJ No. L20, 26.1.88, p. 28.
(**d**) OJ No. L134, 29.5.91, p. 51.
(**e**) OJ No. L248, 23.9.94, p. 22.
(**f**) 1974 c.37; section 15 was amended by the Employment Protection Act 1975 (c.71), Schedule 15, paragraph 6. There are other amendments to section 15 which are not relevant to this provision.
(**g**) S.I. 1956/1771.
(**h**) S.I. 1976/2046.
(**i**) S.I. 1980/1705.
(**j**) S.I. 1992/2932.

(2) These Regulations shall have effect for the purposes of the enforcement of regulation 10 of the Provision and Use of Work Equipment Regulations 1992 as if the addition of the reference to the ATEX Directive in Schedule 1, effected by paragraph (1) above, had been made under section 15 of the Health and Safety at Work etc. Act 1974.

Ian Taylor,
Parliamentary Under-Secretary of State for Science and Technology,
1st February 1996   Department of Trade and Industry

SCHEDULE 1   Regulation 1(2)

REVOCATION OF REGULATIONS

The Electrical Equipment for Explosive Atmospheres (Certification) Regulations 1990**(a)**	1st July 2003
The Electrical Equipment for Explosive Atmospheres (Certification) (Amendment) Regulations 1990**(b)**	1st July 2003
The Electrical Equipment for Explosive Atmospheres (Certification) (Amendment) (No. 2) Regulations 1991**(c)**	1st July 2003
The Electrical Equipment for Explosive Atmospheres (Certification) (Amendment) Regulations 1995**(d)**	1st July 2003
In the Coal and Other Mines (Locomotives) Regulations 1956**(e)**–	1st July 2003
(a) paragraphs (1) and (2) of regulation 3; and	
(b) regulations 4 and 5	
In the Regulations specified in the first column of Schedule 15, the regulations respectively specified in the third column of that Schedule	1st July 2003

SCHEDULE 2   Regulations 2(2) and 6(2)(c)

THE CE MARKING AND OTHER INSCRIPTIONS

1. The CE conformity marking shall consist of the initials 'CE' taking the following form:

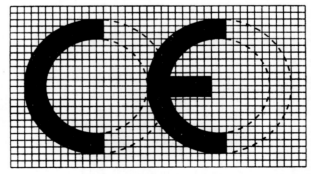

2. If the marking is reduced or enlarged, the proportions given in the above graduated drawing must be respected.

(a) S.I. 1990/13.
(b) S.I. 1990/2377.
(c) S.I. 1991/2826.
(d) S.I. 1995/1186.
(e) S.I. 1956/1771.

3. The various components of the CE marking must have substantially the same vertical dimension, which may not be less than 5 mm.

This minimum dimension may be waived for small-scale equipment, protective systems or devices referred to in Article 1(2).

4. The CE marking shall be followed by the identification number of the notified body where such body is involved in the production control stage.

5. The CE marking shall be affixed distinctly, visibly, legibly and indelibly to equipment and protective systems, supplementary to the provisions of point 1.0.5 of Annex II**(a)**.

6. Subject to paragraph 7 below, where equipment, a protective system or device is the subject of other Community Directives covering other aspects and which also provide for the affixing of the CE marking, such marking shall indicate that the equipment, protective system or device in question is also presumed to conform with the provisions of those other Directives.

7. Where one or more of the other Directives referred to in paragraph 6 above allow the manufacturer, during a transitional period, to choose which arrangements to apply, the CE marking shall indicate conformity only with the Directives applied by the manufacturer. In this case, particulars of the Directives applied, as published in the Official Journal of the European Communities, must be given in the documents, notices or instructions required by the Directives and accompanying such equipment, protective system or device.

8. Where equipment or protective systems are designed for a particular explosive atmosphere, they must be marked accordingly.

9. The affixing of markings on the equipment or protective systems which are likely to deceive third parties as to the meaning and form of the CE marking shall be prohibited. Any other marking may be affixed to the equipment or protective systems, provided that the visibility and legibility of the CE marking is not thereby reduced.

SCHEDULE 3   Regulation 2(2)

ESSENTIAL HEALTH AND SAFETY REQUIREMENTS RELATING TO THE DESIGN AND CONSTRUCTION OF EQUIPMENT AND PROTECTIVE SYSTEMS INTENDED FOR USE IN POTENTIALLY EXPLOSIVE ATMOSPHERES

(Annex II of the ATEX Directive)

Preliminary observations

A. Technological knowledge, which can change rapidly, must be taken into account as far as possible and be utilized immediately.

B. For the devices referred to in Article 1(2), the essential requirements shall apply only in so far as they are necessary for the safe and reliable functioning and operation of those devices with respect to the risks of explosion.

1. COMMON REQUIREMENTS FOR EQUIPMENT AND PROTECTIVE SYSTEMS

1.0. **General requirements**

1.0.1. *Principles of integrated explosion safety*

Equipment and protective systems intended for use in potentially explosive atmospheres must be designed from the point of view of integrated explosion safety.

In this connection, the manufacturer must take measures:

— above all, if possible, to prevent the formation of explosive atmospheres which may be produced or released by equipment and by protective systems themselves,

— to prevent the ignition of explosive atmospheres, taking into account the nature of every electrical and non-electrical source of ignition,

— should an explosion nevertheless occur which could directly or indirectly endanger persons

---

(a) This is a reference to point 1.0.5 of the Essential Health and Safety Requirements contained in Annex II (of the ATEX Directive) which is set out in Schedule 3.

and, as the case may be, domestic animals or property, to halt it immediately and/or to limit the range of explosion flames and explosion pressures to a sufficient level of safety.

1.0.2. Equipment and protective systems must be designed and manufactured after due analysis of possible operating faults in order as far as possible to preclude dangerous situations.

Any misuse which can reasonably be anticipated must be taken into account.

1.0.3. *Special checking and maintenance conditions*

Equipment and protective systems subject to special checking and maintenance conditions must be designed and constructed with such conditions in mind.

1.0.4. *Surrounding area conditions*

Equipment and protective systems must be so designed and constructed as to be capable of coping with actual or foreseeable surrounding area conditions.

1.0.5. *Marking*

All equipment and protective systems must be marked legibly and indelibly with the following minimum particulars;
- name and address of the manufacturer,
- CE marking (see Annex X, point A),
- designation of series or type,
- serial number, if any,
- year of construction,
- the specific marking of explosion protection $\langle\varepsilon_x\rangle$ followed by the symbol of the equipment group and category,
- for equipment-group II, the letter 'G' (concerning explosive atmospheres caused by gases, vapours or mists),
  and/or
  the letter 'D' (concerning explosive atmospheres caused by dust).

Furthermore, where necessary, they must also be marked with all information essential to their safe use.

1.0.6. *Instructions*

(a) All equipment and protective systems must be accompanied by instructions, including at least the following particulars:
   - a recapitulation of the information with which the equipment or protective system is marked, except for the serial number (see 1.0.5.), together with any appropriate additional information to facilitate maintenance (e.g. address of the importer, repairer, etc.);
   - instructions for safe:
     - putting into service,
     - use,
     - assembling and dismantling,
     - maintenance (servicing and emergency repair),
     - installation,
     - adjustment;
   - where necessary, an indication of the danger areas in front of pressure-relief devices;
   - where necessary, training instructions;
   - details which allow a decision to be taken beyond any doubt as to whether an item of equipment in a specific category or a protective system can be used safely in the intended area under the expected operating conditions;
   - electrical and pressure parameters, maximum surface temperatures and other limit values;
   - where necessary, special conditions of use, including particulars of possible misuse which experience has shown might occur;
   - where necessary, the essential characteristics of tools which may be fitted to the equipment or protective system.

(b) The instructions must be drawn up in one of the Community languages by the manufacturer or his authorized representative established in the Community.

On being put into service, all equipment and protective systems must be accompanied by a translation of the instructions in the language or languages of the country in which the equipment or protective system is to be used and by the instructions in the original language.

This translation must be made by either the manufacturer or his authorized representative established in the Community or the person introducing the equipment or protective system into the language area in question.

By way of derogation from this requirement, the maintenance instructions for use by the specialist personnel employed by the manufacturer or his authorized representative established in the Community may be drawn up in a single Community language understood by that personnel.

(c) The instructions must contain the drawings and diagrams necessary for the putting into service, maintenance, inspection, checking of correct operation and, where appropriate, repair of the equipment or protective system, together with all useful instructions, in particular with regard to safety.

(d) Literature describing the equipment or protective system must not contradict the instructions with regard to safety aspects.

1.1. Selection of materials

1.1.1. The materials used for the construction of equipment and protective systems must not trigger off an explosion, taking into account foreseeable operational stresses.

1.1.2. Within the limits of the operating conditions laid down by the manufacturer, it must not be possible for a reaction to take place between the materials used and the constituents of the potentially explosive atmosphere which could impair explosion protection.

1.1.3. Materials must be so selected that predictable changes in their characteristics and their compatibility in combination with other materials will not lead to a reduction in the protection afforded; in particular, due account must be taken of the material's corrosion and wear resistance, electrical conductivity, impact strength, ageing resistance and the effects of temperature variations.

1.2. Design and Construction

1.2.1. Equipment and protective systems must be designed and constructed with due regard to technological knowledge of explosion protection so that they can be safely operated throughout their foreseeable lifetime.

1.2.2. Components to be incorporated into or used as replacements in equipment and protective systems must be so designed and constructed that they function safely for their intended purpose of explosion protection when they are installed in accordance with the manufacturer's instructions.

1.2.3. *Enclosed structures and prevention of leaks*

Equipment which may release flammable gases or dusts must wherever possible employ enclosed structures only.

If equipment contains openings or non-tight joints, these must as far as possible be designed in such a way that developing gases or dusts cannot give rise to explosive atmospheres outside the equipment.

Points where materials are introduced or drawn off must, as far as possible, be designed and equipped so as to limit escapes of flammable materials during filling or draining.

1.2.4. *Dust deposits*

Equipment and protective systems which are intended to be used in areas exposed to dust must be so designed that deposit dust on their surfaces is not ignited.

In general, dust deposits must be limited where possible. Equipment and protective systems must be easily cleanable.

The surface temperatures of equipment parts must be kept well below the glow temperature of the deposit dust.

The thickness of deposit dust must be taken into consideration and, if appropriate, means must be taken to limit the temperature in order to prevent a heat build up.

1.2.5. *Additional means of protection*

Equipment and protective systems which may be exposed to certain types of external stresses must be equipped, where necessary, with additional means of protection.

Equipment must withstand relevant stresses, without adverse effect on explosion protection.

17

1.2.6. *Safe opening*

If equipment and protective systems are in a housing or a locked container forming part of the explosion protection itself, it must be possible to open such housing or container only with a special tool or by means of appropriate protection measures.

1.2.7. *Protection against other hazards*

Equipment and protective systems must be so designed and manufactured as to:

(a) avoid physical injury or other harm which might be caused by direct or indirect contact;

(b) assure that surface temperatures of accessible parts or radiation which would cause a danger, are not produced;

(c) eliminate non-electrical dangers which are revealed by experience;

(d) assure that foreseeable conditions of overload shall not give rise to dangerous situations.

Where, for equipment and protective systems, the risks referred to in this paragraph are wholly or partly covered by other Community Directives, this Directive shall not apply or shall cease to apply in the case of such equipment and protective systems and of such risks upon application of those specific Directives.

1.2.8. *Overloading of equipment*

Dangerous overloading of equipment must be prevented at the design stage by means of integrated measurement, regulation and control devices, such as over-current cut-off switches, temperature limiters, differential pressure switches, flowmeters, time-lag relays, overspeed monitors and/or similar types of monitoring devices.

1.2.9. *Flameproof enclosure systems*

If parts which can ignite an explosive atmosphere are placed in an enclosure, measures must be taken to ensure that the enclosure withstands the pressure developed during an internal explosion of an explosive mixture and prevents the transmission of the explosion to the explosive atmosphere surrounding the enclosure.

1.3. Potential ignition sources

1.3.1. *Hazards arising from different ignition sources*

Potential ignition sources such as sparks, flames, electric arcs, high surface temperatures, acoustic energy, optical radiation, electromagnetic waves and other ignition sources must not occur.

1.3.2. *Hazards arising from static electricity*

Electrostatic charges capable of resulting in dangerous discharges must be prevented by means of appropriate measures.

1.3.3. *Hazards arising from stray electric and leakage currents*

Stray electric and leakage currents in conductive equipment parts which could result in, for example, the occurrence of dangerous corrosion, overheating of surfaces or sparks capable of provoking an ignition must be prevented.

1.3.4. *Hazards arising from overheating*

Overheating caused by friction or impacts occurring, for example, between materials and parts in contact with each other while rotating or through the intrusion of foreign bodies must, as far as possible, be prevented at the design stage.

1.3.5. *Hazards arising from pressure compensation operations*

Equipment and protective systems must be so designed or fitted with integrated measuring, control and regulation devices that pressure compensations arising from them do not generate shock waves or compressions which may cause ignition.

1.4. Hazards arising from external effects

1.4.1. Equipment and protective systems must be so designed and constructed as to be capable of performing their intended function in full safety, even in changing environmental conditions and in the presence of extraneous voltages, humidity, vibrations, contamination and other external effects, taking into account the limits of the operating conditions established by the manufacturer.

1.4.2. Equipment parts used must be appropriate to the intended mechanical and thermal stresses and capable of withstanding attack by existing or foreseeable aggressive substances.

1.5. **Requirements in respect of safety-related devices**

1.5.1. Safety devices must function independently of any measurement or control devices required for operation.

As far as possible, failure of a safety device must be detected sufficiently rapidly by appropriate technical means to ensure that there is only very little likelihood that dangerous situations will occur.

For electrical circuits the fail-safe principle is to be applied in general.

Safety-related switching must in general directly actuate the relevant control devices without intermediate software command.

1.5.2. In the event of a safety device failure, equipment and/or protective systems shall, wherever possible, be secured.

1.5.3. Emergency stop controls of safety devices must, as far as possible, be fitted with restart lockouts. A new start command may take effect on normal operation only after the restart lockouts have been intentionally reset.

1.5.4. *Control and display units*

Where control and display units are used, they must be designed in accordance with ergonomic principles in order to achieve the highest possible level of operating safety with regard to the risk of explosion.

1.5.5. *Requirements in respect of devices with a measuring function for explosion protection*

In so far as they relate to equipment used in explosive atmospheres, devices with a measuring function must be designed and constructed so that they can cope with foreseeable operating requirements and special conditions of use.

1.5.6. Where necessary, it must be possible to check the reading accuracy and serviceability of devices with a measuring function.

1.5.7. The design of devices with a measuring function must incorporate a safety factor which ensures that the alarm threshold lies far enough outside the explosion and/or ignition limits of the atmospheres to be registered, taking into account, in particular, the operating conditions of the installation and possible aberrations in the measuring system.

1.5.8. *Risks arising from software*

In the design of software-controlled equipment, protective systems and safety devices, special account must be taken of the risks arising from faults in the programme.

1.6. **Integration of safety requirements relating to the system**

1.6.1. Manual override must be possible in order to shut down the equipment and protective systems incorporated within automatic processes which deviate from the intended operating conditions, provided that this does not compromise safety.

1.6.2. When the emergency shutdown system is actuated, accumulated energy must be dispersed as quickly and as safely as possible or isolated so that it no longer constitutes a hazard.

This does not apply to electrochemically-stored energy.

1.6.3. *Hazards arising from power failure*

Where equipment and protective systems can give rise to a spread of additional risks in the event of a power failure, it must be possible to maintain them in a safe state of operation independently of the rest of the installation.

1.6.4. *Hazards arising from connections*

Equipment and protective systems must be fitted with suitable cable and conduit entries.

When equipment and protective systems are intended for use in combination with other equipment and protective systems, the interface must be safe.

1.6.5. *Placing of warning devices as parts of equipment*

Where equipment or protective systems are fitted with detection or alarm devices for monitoring the occurrence of explosive atmospheres, the necessary instructions must be provided to enable them to be provided at the appropriate places.

2. SUPPLEMENTARY REQUIREMENTS IN RESPECT OF EQUIPMENT

2.0. Requirements applicable to equipment in category M of equipment-group I

2.0.1. Requirements applicable to equipment in category M 1 of equipment-group I

2.0.1.1. Equipment must be so designed and constructed that sources of ignition do not become active, even in the event of rare incidents relating to equipment.

Equipment must be equipped with means of protection such that:

— either, in the event of failure of one means of protection, at least an independent second means provides the requisite level of protection,

— or, the requisite level of protection is ensured in the event of two faults occurring independently of each other.

Where necessary, this equipment must be equipped with additional special means of protection.

It must remain functional with an explosive atmosphere present.

2.0.1.2. Where necessary, equipment must be so constructed that no dust can penetrate it.

2.0.1.3. The surface temperatures of equipment parts must be kept clearly below the ignition temperature of the foreseeable air/dust mixtures in order to prevent the ignition of suspended dust.

2.0.1.4. Equipment must be so designed that the opening of equipment parts which may be sources of ignition is possible only under non-active or intrinsically safe conditions. Where it is not possible to render equipment non-active, the manufacturer must affix a warning label to the opening part of the equipment.

If necessary, equipment must be fitted with appropriate additional interlocking systems.

2.0.2. Requirements applicable to equipment in category M 2 of equipment-group I

2.0.2.1. Equipment must be equipped with means of protection ensuring that sources of ignition do not become active during normal operation, even under more severe operating conditions, in particular those arising from rough handling and changing environmental conditions.

The equipment is intended to be de-energized in the event of an explosive atmosphere.

2.0.2.2. Equipment must be so designed that the opening of equipment parts which may be sources of ignition is possible only under non-active conditions or via appropriate interlocking systems. Where it is not possible to render equipment non-active, the manufacturer must affix a warning label to the opening part of the equipment.

2.0.2.3. The requirements regarding explosion hazards arising from dust applicable to category M 1 must be applied.

2.1. Requirements applicable to equipment in category 1 of equipment-group II

2.1.1. Explosive atmospheres caused by gases, vapours or hazes

2.1.1.1. Equipment must be so designed and constructed that sources of ignition do not become active, even in event of rare incidents relating to equipment.

It must be equipped with means of protection such that:

— either, in the event of failure of one means of protection, at least an independent second means provides the requisite level of protection,

— or, the requisite level of protection is ensured in the event of two faults occurring independently of each other.

2.1.1.2. For equipment with surfaces which may heat up, measures must be taken to ensure that the stated maximum surface temperatures are not exceeded even in the most unfavourable circumstances.

Temperature rises caused by heat build-ups and chemical reactions must also be taken into account.

2.1.1.3. Equipment must be so designed that the opening of equipment parts which might be sources of ignition is possible only under non-active or intrinsically safe conditions. Where it is not possible to render equipment non-active, the manufacturer must affix a warning label to the opening part of the equipment.

If necessary, equipment must be fitted with appropriate additional interlocking systems.

2.1.2. Explosive atmospheres caused by air/dust mixtures

2.1.2.1. Equipment must be so designed and constructed that ignition of air/dust mixtures does not occur even in the event of rare incidents relating to equipment.

It must be equipped with means of protection such that

— either, in the event of failure of one means of protection, at least an independent second means provides the requisite level of protection,

The Regulatory Reform (Fire Safety) Order 2005

Elimination or reduction of risks from dangerous substances

1.—(1) Where a dangerous substance is present in or on the premises, the responsible person must ensure that risk to relevant persons related to the presence of the substance is either eliminated or reduced so far as is reasonably practicable.

(2) In complying with his duty under paragraph (1), the responsible person must, so far as is reasonably practicable, replace a dangerous substance, or the use of a dangerous substance, with a substance or process which either eliminates or reduces the risk to relevant persons.

(3) Where it is not reasonably practicable to eliminate risk pursuant to paragraphs (1) and (2), the responsible person must, so far as is reasonably practicable, apply measures consistent with the risk assessment and appropriate to the nature of the activity or operation, including the measures specified in Part 4 of Schedule 1 to this Order to—

(a) control the risk, and

(b) mitigate the detrimental effects of a fire.

(4) The responsible person must—

(a) arrange for the safe handling, storage and transport of dangerous substances and waste containing dangerous substances; and

(b) ensure that any conditions necessary pursuant to this Order for ensuring the elimination or reduction of risk are maintained.

Additional emergency measures in respect of dangerous substances

2.—(1) Subject to paragraph (4), in order to safeguard the safety of relevant persons arising from an accident, incident or emergency related to the presence of a dangerous substance in or on the premises, the responsible person must ensure that—

(a) information on emergency arrangements is available, including—

(i) details of relevant work hazards and hazard identification arrangements; and

(ii) specific hazards likely to arise at the time of an accident, incident or emergency;

(b) suitable warning and other communication systems are established to enable an appropriate response, including remedial actions and rescue operations, to be made immediately when such an event occurs;

(c) where necessary, before any explosion conditions are reached, visual or audible warnings are given and relevant persons withdrawn; and

(d) where the risk assessment indicates it is necessary, escape facilities are provided and maintained to ensure that, in the event of danger, relevant persons can leave endangered places promptly and safely.

(2) Subject to paragraph (4), the responsible person must ensure that the information required by article **Error! Reference source not found.**(1)(a) and paragraph (1)(a) of this article, together with information on the matters referred to in paragraph (1)(b) and (d) is—

(a) made available to relevant accident and emergency services to enable those services, whether internal or external to the premises, to prepare their own response procedures and precautionary measures; and

(b) displayed at the premises, unless the results of the risk assessment make this unnecessary.

(3) Subject to paragraph (4), in the event of a fire arising from an accident, incident or emergency related to the presence of a dangerous substance in or on the premises, the responsible person must ensure that—

(a) immediate steps are taken to—

(i) mitigate the effects of the fire;

(ii) restore the situation to normal; and

(iii) inform those relevant persons who may be affected; and

(b) only those persons who are essential for the carrying out of repairs and other necessary work are permitted in the affected area and they are provided with—

(i) appropriate personal protective equipment and protective clothing; and

(ii) any necessary specialised safety equipment and plant,

which must be used until the situation is restored to normal.

(4) Paragraphs (1) to (3) do not apply where—

(a) the results of the risk assessment show that, because of the quantity of each dangerous substance in or on the premises, there is only a slight risk to relevant persons; and

the measures taken by the responsible person to comply with his duty under article 1 are sufficient to control that risk.

— or, the requisite level of protection is ensured in the event of two faults occurring independently of each other.

2.1.2.2. Where necessary, equipment must be so designed that dust can enter or escape from the equipment only at specifically designated points.

This requirement must also be met by cable entries and connecting pieces.

2.1.2.3. The surface temperatures of equipment parts must be kept well below the ignition temperature of the foreseeable air/dust mixtures in order to prevent the ignition of suspended dust.

2.1.2.4. With regard to the safe opening of equipment parts, requirement 2.1.1.3 applies.

2.2. Requirements for category 2 of equipment-group II

2.2.1. *Explosive atmospheres caused by gases, vapours or mists*

2.2.1.1. Equipment must be so designed and constructed as to prevent ignition sources arising, even in the event of frequently occurring disturbances or equipment operating faults, which normally have to be taken into account.

2.2.1.2. Equipment parts must be so designed and constructed that their stated surface temperatures are not exceeded, even in the case of risks arising from abnormal situations anticipated by the manufacturer.

2.2.1.3. Equipment must be so designed that the opening of equipment parts which might be sources of ignition is possible only under non-active conditions or via appropriate interlocking systems. Where it is not possible to render equipment non-active, the manufacturer must affix a warning label to the opening part of the equipment.

2.2.2. *Explosive atmospheres caused by air/dust mixtures*

2.2.2.1. Equipment must be designed and constructed so that ignition of air/dust mixtures is prevented, even in the event of frequently occurring disturbances or equipment operating faults which normally have to be taken into account.

2.2.2.2. With regard to surface temperatures, requirement 2.1.2.3 applies.

2.2.2.3. With regard to protection against dust, requirement 2.1.2.2 applies.

2.2.2.4. With regard to the safe opening of equipment parts, requirement 2.2.1.3 applies.

2.3. Requirements applicable to equipment in category 3 of equipment-group II

2.3.1. *Explosive atmospheres caused by gases, vapours or mists*

2.3.1.1. Equipment must be so designed and constructed as to prevent foreseeable ignition sources which can occur during normal operation.

2.3.1.2. Surface temperatures must not exceed the stated maximum surface temperatures under intended operating conditions. Higher temperatures in exceptional circumstances may be allowed only if the manufacturer adopts special additional protective measures.

2.3.2. *Explosive atmospheres caused by air/dust mixtures*

2.3.2.1. Equipment must be so designed and constructed that air/dust mixtures cannot be ignited by foreseeable ignition sources likely to exist during normal operation.

2.3.2.2. With regard to surface temperatures, requirement 2.1.2.3 applies.

2.3.2.3. Equipment, including cable entries and connecting pieces, must be so constructed that, taking into account the size of its particles, dust can neither develop explosive mixtures with air nor form dangerous accumulations inside the equipment.

3. SUPPLEMENTARY REQUIREMENTS IN RESPECT OF PROTECTIVE SYSTEMS

3.0. General requirements

3.0.1. Protective systems must be dimensioned in such a way as to reduce the effects of an explosion to a sufficent level of safety.

3.0.2. Protective systems must be designed and capable of being positional in such a way that explosions are prevented from spreading through dangerous chain reactions or flashover and incipient explosions do not become detonations.

3.0.3. In the event of a power failure, protective systems must retain their capacity to function for a period sufficient to avoid a dangerous situation.

3.0.4. Protective systems must not fail due to outside interference.

3.1. **Planning and design**

3.1.1. *Characteristics of materials*

With regard to the characteristics of materials, the maximum pressure and temperature to be taken into consideration at the planning stage are the expected pressure during an explosion occurring under extreme operating conditions and the anticipated heating effect of the flame.

3.1.2. Protective systems designed to resist or contain explosions must be capable of withstanding the shock wave produced without losing system integrity.

3.1.3. Accessories connected to protective systems must be capable of withstanding the expected maximum explosion pressure without losing their capacity to function.

3.1.4. The reactions caused by pressure in peripheral equipment and connected pipe-work must be taken into consideration in the planning and design of protective systems.

3.1.5. *Pressure-relief systems*

If it is likely that stresses on protective systems will exceed their structural strength, provision must be made in the design for suitable pressure-relief devices which do not endanger persons in the vicinity.

3.1.6. *Explosion suppression systems*

Explosion suppression systems must be so planned and designed that they react to an incipient explosion at the earliest possible stage in the event of an incident and counteract it to best effect, with due regard to the maximum rate of pressure increase and the maximum explosion pressure.

3.1.7. *Explosion decoupling systems*

Decoupling systems intended to disconnect specific equipment as swiftly as possible in the event of incipient explosions by means of appropriate devices must be planned and designed so as to remain proof against the transmission of internal ignition and to retain their mechanical strength under operating conditions.

3.1.8. Protective systems must be capable of being integrated into a circuit with a suitable alarm threshold so that, if necessary, there is cessation of product feed and output and shutdown of equipment parts which can no longer function safely.

SCHEDULE 4   Regulation 2(2)

(Annex I of the ATEX Directive: references in this Annex to Annex II are references to the Essential Health and Safety Requirements set out in Schedule 3)

CRITERIA DETERMINING THE CLASSIFICATION OF EQUIPMENT-GROUPS INTO CATEGORIES

1. Equipment-group I

(a) Category M 1 comprises equipment designed and, where necessary, equipped with additional special means of protection to be capable of functioning in conformity with the operational parameters established by the manufacturer and ensuring a very high level of protection.

Equipment in this category is intended for use in underground parts of mines as well as those parts of surface installations of such mines endangered by firedamp and/or combustible dust.

Equipment in this category is required to remain functional, even in the event of rare incidents relating to equipment, with an explosive atmosphere present, and is characterized by means of protection such that:

— either, in the event of failure of one means of protection, at least an independent second means provides the requisite level of protection,

— or the requisite level of protection is assured in the event of two faults occurring independently of each other.

Equipment in this category must comply with the supplementary requirements referred to in Annex II, 2.0.1.

(b) Category M 2 comprises equipment designed to be capable of functioning in conformity with the operational parameters established by the manufacturer and ensuring a high level of protection.

Equipment in this category is intended for use in underground parts of mines as well as those parts of surface installations of such mines likely to be endangered by firedamp and/or combustible dust.

This equipment is intended to be de-energized in the event of an explosive atmosphere.

The means of protection relating to equipment in this category assure the requisite level of protection during normal operation and also in the case of more severe operating conditions, in particular those arising from rough handling and changing environmental conditions.

Equipment in this category must comply with the supplementary requirements referred to in Annex II, 2.0.2.

2. Equipment-group II

(a) Category 1 comprises equipment designed to be capable of functioning in conformity with the operational parameters established by the manufacturer and ensuring a very high level of protection.

Equipment in this category is intended for use in areas in which explosive atmospheres caused by mixtures of air and gases, vapours or mists or by air/dust mixtures are present continuously, for long periods or frequently.

Equipment in this category must ensure the requisite level of protection, even in the event of rare incidents relating to equipment, and is characterized by means of protection such that:

— either, in the event of failure of one means of protection, at least an independent second means provides the requisite level of protection,

— or the requisite level of protection is assured in the event of two faults occurring independently of each other.

Equipment in this category must comply with the supplementary requirements referred to in Annex II, 2.1.

(b) Category 2 comprises equipment designed to be capable of functioning in conformity with the operational parameters established by the manufacturer and of ensuring a high level of protection.

Equipment in this category is intended for use in areas in which explosive atmospheres caused by gases, vapours, mists or air/dust mixtures are likely to occur.

The means of protection relating to equipment in this category ensure the requisite level of protection, even in the event of frequently occurring disturbances or equipment faults which normally have to be taken into account.

Equipment in this category must comply with the supplementary requirements referred to in Annex II, 2.2.

(c) Category 3 comprises equipment designed to be capable of functioning in conformity with the operating parameters established by the manufacturer and ensuring a normal level of protection.

Equipment in this category is intended for use in areas in which explosive atmospheres caused by gases, vapours, mists, or air/dust mixtures are unlikely to occur or, if they do occur, are likely to do so only infrequently and for a short period only.

Equipment in this category ensures the requisite level of protection during normal operation.

Equipment in this category must comply with the supplementary requirements referred to in Annex II, 2.3.

SCHEDULE 5
Regulation 4

EXCLUDED EQUIPMENT, PROTECTIVE SYSTEMS AND DEVICES

Medical devices intended for use in a medical environment,

Equipment and protective systems where the explosion hazard results exclusively from the presence of explosive substances or unstable chemical substances,

Equipment intended for use in domestic and non-commercial environments where potentially explosive atmospheres may only rarely be created, solely as a result of the accidental leakage of fuel gas,

Personal protective equipment covered by Directive 89/686/EEC,

Seagoing vessels and mobile offshore units together with equipment on board such vessels or units,

Means of transport, i.e. vehicles and their trailers intended solely for transporting passengers by air or by road, rail or water networks, as well as means of transport in so far as such means are designed for transporting goods by air, by public road or rail networks or by water. Vehicles intended for use in a potentially explosive atmosphere shall not be excluded,

The equipment covered by Article 223(1)(b) of the Treaty establishing the European Community.

(Annex III of the ATEX Directive)

MODULE EC-TYPE EXAMINATION

1. This module describes that part of the procedure by which a notified body ascertains and attests that a specimen representative of the production envisaged meets the relevant applicable provisions of the Directive.

2. The application for the EC-type examination shall be lodged by the manufacturer or his authorized representative established within the Community with a notified body of his choice.

The application shall include:
— the name and address of the manufacturer and, if the application is lodged by the authorized representative, his name and address in addition;
— a written declaration that the same application has not been lodged with any other notified body;
— the technical documentation, as described in point 3.

The applicant shall place at the disposal of the notified body a specimen representative of the production envisaged and hereinafter called 'type'. The notified body may request further specimens if needed for carrying out the test programme.

3. The technical documentation shall enable the conformity of the product with the requirements of the Directive to be assessed. It shall, to the extent necessary for such assessment, cover the design, manufacture and operation of the product and shall to that extent contain:
— a general type-description;
— design and manufacturing drawings and layouts of components, sub-assemblies, circuits, etc.;
— descriptions and explanations necessary for the understanding of said drawings and layouts and the operation of the product;
— a list of the standards referred to in Article 5, applied in full or in part, and descriptions of the solutions adopted to meet the essential requirements of the Directive where the standards referred to in Article 5 have not been applied;
— results of design calculations made, examinations carried out, etc.;
— test reports.

4. The notified body shall:

4.1. examine the technical documentation, verify that the type has been manufactured in conformity with the technical documentation and identify the elements which have been designed in accordance with the relevant provisions of the standards referred to in Article 5, as well as the components which have been designed without applying the relevant provisions of those standards;

4.2. perform or have performed the appropriate examinations and necessary tests to check whether the solutions adopted by the manufacturer meet the essential requirements of the Directive where the standards referred to in Article 5 have not been applied;

4.3. perform or have performed the appropriate examinations and necessary tests to check whether these have actually been applied, where the manufacturer has chosen to apply the relevant standards;

4.4. agree with the applicant the location where the examinations and necessary tests shall be carried out.

5. Where the type meets the provisions of the Directive, the notified body shall issue an EC-type-examination certificate to the applicant. The certificate shall contain the name and address of the manufacturer, conclusions of the examination and the necessary data for identification of the approved type.

A list of the relevant parts of the technical documentation shall be annexed to the certificate and a copy kept by the notified body.

If the manufacturer or his authorized representative established in the Community is denied a type certification, the notified body shall provide detailed reasons for such denial.

Provision shall be made for an appeals procedure.

6. The applicant shall inform the notified body which holds the technical documentation concerning the EC-type-examination certificate of all modifications to the approved equipment or protective system which must receive further approval where such changes may affect conformity with the essential requirements or with the prescribed conditions for use of the product. This further approval is given in the form of an addition to the original EC-type-examination certificate.

7. Each notified body shall communicate to the other notified bodies the relevant information concerning the EC-type-examination certificates and additions issued and withdrawn.

8. The other notified bodies may receive copies of the EC-type-examination certificates and/or their additions. The annexes to the certificates shall be kept at the disposal of the other notified bodies.

9. The manufacturer or his authorized representative established in the Community shall keep with the technical documentation copies of EC-type-examination certificates and their additions for a period ending at least 10 years after the last equipment or protective system was manufactured.

Where neither the manufacturer nor his authorized representative is established within the Community, the obligation to keep the technical documentation available shall be the responsibility of the person who places the product on the Community market.

SCHEDULE 7   Regulation 10

(Annex IV of the ATEX Directive)

MODULE: PRODUCTION QUALITY ASSURANCE

1. This module describes the procedure whereby the manufacturer who satisfies the obligations of point 2 ensures and declares that the products concerned are in conformity with the type as described in the EC-type-examination certificate and satisfy the requirements of the Directive which apply to them. The manufacturer, or his authorized representative established in the Community, shall affix the CE marking to each piece of equipment and draw up a written declaration of conformity. The CE marking shall be accompanied by the identification number of the notified body responsible for EC monitoring, as specified in Section 4.

2. The manufacturer shall operate an approved quality system for production, final equipment inspection and testing as specified in Section 3 and shall be subject to monitoring as specified in Section 4.

3. Quality system

3.1. The manufacturer shall lodge an application for assessment of his quality system with a notified body of his choice, for the equipment concerned.

The application shall include:
— all relevant information for the product category envisaged;
— the documentation concerning the quality system;
— technical documentation on the approved type and a copy of the EC-type-examination certificate.

3.2. The quality system shall ensure compliance of the equipment with the type as described in the EC-type-examination certificate and with the requirements of the Directive which apply to them.

All the elements, requirements and provisions adopted by the manufacturer shall be documented in a systematic and orderly manner in the form of written policies, procedures and instructions. The quality system documentation must permit a consistent interpretation of quality programmes, plans, manuals and records.

It shall contain, in particular, an adequate description of
— the quality objectives and the organizational structure, responsibilities and powers of the management with regard to equipment quality;
— the manufacturing, quality control and quality assurance techniques, processes and systematic actions which will be used;

- the examinations and tests which will be carried out before, during and after manufacture and the frequency with which they will be carried out;
- the quality records, such as inspection reports and test data, calibration data, reports on the qualifications of the personnel concerned, etc.;
- the means to monitor the achievement of the required equipment quality and the effective operation of the quality system.

3.3. The notified body shall assess the quality system to determine whether it satisfies the requirements referred to in Section 3.2. It shall presume conformity with these requirements in respect of quality systems which implement the relevant harmonized standard. The auditing team shall have at least one member with experience of evaluation in the equipment technology concerned. The evaluation procedure shall include an inspection visit to the manufacturer's premises. The decision shall be notified to the manufacturer. The notification shall contain the conclusions of the examination and the reasoned assessment decision.

3.4. The manufacturer shall undertake to fulfil the obligations arising out of the quality system as approved and to uphold the system so that it remains adequate and efficient.

The manufacturer or his authorized representative shall inform the notified body which has approved the quality system of any intended updating of the quality system.

The notified body shall evaluate the modifications proposed and decide whether the amended quality system will still satisfy the requirements referred to in Section 3.2 or whether a re-assessment is required.

It shall notify its decision to the manufacturer. The notification shall contain the conclusions of the examination and the reasoned assessment decision.

4. Surveillance under the responsibility of the notified body

4.1. The purpose of surveillance is to make sure that the manufacturer duly fulfils the obligations arising out of the approved quality system.

4.2. The manufacturer shall, for inspection purposes, allow the notified body access to the manufacture, inspection, testing and storage premises and shall provide it with all necessary information, in particular
- the quality system documentation
- the quality records, such as inspection reports and text data, calibration data, reports on the qualifications of the personnel concerned, etc.

4.3. The notified body shall periodically carry out audits to ensure that the manufacturer maintains and applies the quality system and shall provide an audit report to the manufacturer.

4.4. Furthermore, the notified body may pay unexpected visits to the manufacturer. During such visits, the notified body may carry out tests, or arrange for tests to be carried out, to check that the quality system is functioning correctly, if necessary. The notified body shall provide the manufacturer with a visit report and, if a test has taken place, with a test report.

5. The manufacturer shall, for a period ending at least 10 years after the last piece of equipment was manufactured, keep at the disposal of the national authorities:
- the documentation referred to in the second indent of Section 3.1;
- the updating referred to in the second paragraph of Section 3.4;
- the decisions and reports from the notified body which are referred to in Section 3.4, last paragraph, Section 4.3 and Section 4.4.

6. Each notified body shall apprise the other notified bodies of the relevant information concerning the quality system approvals issued and withdrawn.

(Annex V of the ATEX Directive)

MODULE: PRODUCT VERIFICATION

1.   This module describes the procedure whereby a manufacturer or his authorized representative established within the Community checks and attests that the equipment subject to the provisions of point 3 are in conformity with the type as described in the EC-type-examination certificate and satisfy the relevant requirements of the Directive.

2.   The manufacturer shall take all measures necessary to ensure that the manufacturing process guarantees conformity of the equipment with the type as described in the EC-type-examination certificate and with the requirements of the Directive which apply to them. The manufacturer or his authorized representative established in the Community shall affix the CE marking to each piece of equipment and shall draw up a declaration of conformity.

3.   The notified body shall carry out the appropriate examinations and tests in order to check the conformity of the equipment, protective system or device referred to in Article 1(2), with the relevant requirements of the Directive, by examining and testing every product as specified in Section 4.

The manufacturer or his authorized representative shall keep a copy of the declaration of conformity for a period ending at least 10 years after the last piece of equipment was manufactured.

4. Verification by examination and testing of each piece of equipment

4.1.   All equipment shall be individually examined and appropriate tests as set out in the relevant standard(s) referred to in Article 5 or equipment tests shall be carried out in order to verify their conformity with the type as described in the EC-type-examination certificate and the relevant requirements of the Directive.

4.2.   The notified body shall affix or have affixed its identification number to each approved item of equipment and shall draw up a written certificate of conformity relating to the tests carried out.

4.3.   The manufacturer or his authorized representative shall ensure that he is able to supply the notified body's certificates of conformity on request.

(Annex VI of the ATEX Directive)

MODULE: CONFORMITY TO TYPE

1.   This module describes that part of the procedure whereby the manufacturer or his authorized representative established within the Community ensures and declares that the equipment in question is in conformity with the type as described in the EC-type-examination certificate and satisfy the requirements of the Directive applicable to them. The manufacturer or his authorized representative established within the Community shall affix the CE marking to each piece of equipment and draw up a written declaration of conformity.

2.   The manufacturer shall take all measures necessary to ensure that the manufacturing process assures compliance of the manufactured equipment or protective systems with the type as described in the EC-type-examination certificate and with the relevant requirements of the Directive.

3.   The manufacturer or his authorised representative shall keep a copy of the declaration of conformity for a period ending at least 10 years after the last piece of equipment was manufactured. Where neither the manufacturer nor his authorized representative is established within the Community, the obligation to keep the technical documentation available shall be the responsibility of the person who places the equipment or protective system on the Community market.

For each piece of equipment manufactured, tests relating to the anti-explosive protection aspects of the product shall be carried out by the manufacturer or on his behalf. The tests shall be carried out under the responsibility of a notified body, chosen by the manufacturer.

On the responsibility of the notified body, the manufacturer shall affix the former's identification number during the manufacturing process.

(Annex VII of the ATEX Directive)

MODULE: PRODUCT QUALITY ASSURANCE

1. This module describes the procedure whereby the manufacturer who satisfies the obligations of Section 2 ensures and declares that the equipment is in conformity with the type as described in the EC-type-examination certificate. The manufacturer or his authorized representative established within the Community shall affix the CE marking to each product and draw up a written declaration of conformity. The CE marking shall be accompanied by the identification number of the notified body responsible for surveillance as specified in Section 4.

2. The manufacturer shall operate an approved quality system for the final inspection and testing of equipment as specified in Section 3 below and shall be subject to surveillance as specified in Section 4 below.

3. Quality system

3.1. The manufacturer shall lodge an application for assessment of his quality system for the equipment and protective systems, with a notified body of his choice.

The application shall include:
— all relevant information for the product category envisaged;
— documentation on the quality system;
— technical documentation on the approved type and a copy of the EC-type-examination certificate.

3.2. Under the quality system, each piece of equipment shall be examined and appropriate tests as set out in the relevant standard(s) referred to in Article 5 or equivalent tests shall be carried out in order to ensure its conformity with the relevant requirements of the Directive. All the elements, requirements and provisions adopted by the manufacturer shall be documented in a systematic and orderly manner in the form of written policies, procedures and instruments. This quality system documentation must permit a consistent interpretation of the quality programmes, plans, manuals and records.

It shall contain, in particular, an adequate description of:
— the quality objectives and the organizational structure, responsibilities and powers of the management with regard to product quality;
— the examinations and tests which will be carried out after manufacture;
— the means to monitor the effective operation of the quality system;
— quality records, such as inspection reports and test data, calibration data, reports on the qualifications of the personnel concerned, etc.

3.3. The notified body shall assess the quality system to determine whether it satisfies the requirements referred to in Section 3.2. It shall presume conformity with these requirements in respect of quality systems which implement the relevant harmonized standard.

The auditing team shall have at least one member experienced as an assessor in the product technology concerned. The assessment procedure shall include an assessment visit to the manufacturer's premises.

The decision shall be notified to the manufacturer. The notification shall contain the conclusions of the examination and the reasoned assessment decision.

3.4. The manufacturer shall undertake to discharge the obligations arising from the quality system as approved and to maintain it in an appropriate and efficient manner.

The manufacturer or his authorized representative shall inform the notified body which has approved the quality system of any intended updating of the quality system.

The notified body shall evaluate the modifications proposed and decide whether the modified quality system will still satisfy the requirements referred to in Section 3.2 or whether a re-assessment is required.

It shall notify its decision to the manufacturer. The notification shall contain the conclusions of the examination and the reasoned assessment decision.

4. Surveillance under the responsibility of the notified body

4.1. The purpose of surveillance is to ensure that the manufacturer duly fulfils the obligations arising out of the approved quality system.

4.2. The manufacturer shall for inspection purposes allow the notified body access to the inspection, testing and storage premises and shall provide it with all necessary information, in particular:

— quality system documentation;
— technical documentation;
— quality records, such as inspection reports and test data, calibration data, reports on the qualifications of the personnel concerned, etc.

4.3. The notified body shall periodically carry out audits to ensure that the manufacturer maintains and applies the quality system and shall provide an audit report to the manufacturer.

4.4. Furthermore, the notified body may pay unexpected visits to the manufacturer. At the time of such visits, the notified body may carry out tests or arrange for tests to be carried out in order to check the proper functioning of the quality system, where necessary; it shall provide the manufacturer with a visit report and, if a test has been carried out, with a test report.

5. The manufacturer shall, for a period ending at least 10 years after the last piece of equipment was manufactured, keep at the disposal of the national authorities:

— the documentation referred to in the third indent of Section 3.1;
— the updating referred to in the second paragraph of Section 3.4;
— the decisions and reports from the notified body which are referred to in Section 3.4, last paragraph, Section 4.3 and Section 4.4.

6. Each notified body shall forward to the other notified bodies the relevant information concerning the quality system approvals issued and withdrawn.

SCHEDULE 11   Regulation 10

(Annex VIII of the ATEX Directive)

MODULE: INTERNAL CONTROL OF PRODUCTION

1. This module describes the procedure whereby the manufacturer or his authorized representative established within the Community, who carries out the obligations laid down in Section 2, ensures and declares that the equipment satisfy the requirements of the Directive applicable to it. The manufacturer or his authorized representative established within the Community shall affix the CE marking to each piece of equipment and draw up a written declaration of conformity.

2. The manufacturer shall establish the technical documentation described in Section 3 and he or his authorized representative established within the Community shall keep it at the disposal of the relevant national authorities for inspection purposes for a period ending at least 10 years after the last piece of equipment was manufactured.

Where neither the manufacturer nor his authorized representative is established within the Community, the obligation to keep the technical documentation available shall be the responsibility of the person who places the equipment on the Community market.

3. Technical documentation shall enable the conformity of the equipment with the relevant requirements of the Directive to be assessed. It shall, to the extent necessary for such assessment, cover the design, manufacture and operation of the product. It shall contain:

— a general description of the equipment,
— conceptual design and manufacturing drawings and schemes of components, sub-assemblies, circuits, etc.,
— descriptions and explanations necessary for the understanding of said drawings and schemes and the operation of the equipment,
— a list of the standards applied in full or in part, and descriptions of the solutions adopted to meet the safety aspects of the Directive where the standards have not been applied,
— results of design calculations made, examinations carried out, etc.,
— test reports.

4. The manufacturer or his authorized representative shall keep a copy of the declaration of conformity with the technical documentation.

5. The manufacturer shall take all measures necessary to ensure that the manufacturing process guarantees compliance of the manufactured equipment with the technical documentation referred to in Section 2 and with the requirements of the Directive applicable to such equipment.

(Annex IX of the ATEX Directive)

MODULE: UNIT VERIFICATION

1. This module describes the procedure whereby the manufacturer ensures and declares that the equipment or protective system which has been issued with the certificate referred to in Section 2 conforms to the requirements of the Directive which are applicable to it. The manufacturer or his authorized representative in the Community shall affix the CE marking to the equipment or protective system and draw up a declaration of conformity.

2. The notified body shall examine the individual equipment or protective system and carry out the appropriate tests as set out in the relevant standard(s) referred to in Article 5, or equivalent tests, to ensure its conformity with the relevant requirements of the Directive.

The notified body shall affix, or cause to be affixed, its identification number on the approved equipment or protective system and shall draw up a certificate of conformity concerning the tests carried out.

3. The aim of the technical documentation is to enable conformity with the requirements of the Directive to be assessed and the design, manufacture and operation of the equipment or protective system to be understood.

The documentation shall contain:
— a general description of the product;
— conceptual design and manufacturing drawings and layouts of components, sub-assemblies, circuits, etc.;
— descriptions and explanations necessary for the understanding of said drawings and layouts and the operation of the equipment or protective system;
— a list of the standards referred to in Article 5, applied in full or in part, and descriptions of the solutions adopted to meet the essential requirements of the Directive where the standards referred to in Article 5 have not been applied;
— results of design calculations made, examinations carried out, etc.;
— test reports.

SCHEDULE 13   Regulation 14(1)(a)

(Part B of Annex X of the ATEX Directive)

(Content of the EC declaration of conformity)

The EC declaration of conformity must contain the following elements:
— the name or identification mark and the address of the manufacturer or his authorized representative established within the Community;
— a description of the equipment, protective system, or device referred to in Article 1(2);
— all relevant provisions fulfilled by the equipment, protective system, or device referred to in Article 1(2);
— where appropriate, the name, identification number and address of the notified body and the number of the EC-type-examination certificate;
— where appropriate, reference to the harmonized standards;
— where appropriate, the standards and technical specifications which have been used;
— where appropriate, references to other Community Directives which have been applied;
— identification of the signatory who has been empowered to enter into commitments on behalf of the manufacturer or his authorized representative established within the Community.

ENFORCEMENT

Enforcement in relation to relevant products

1. In relation to relevant products–

    (a) it shall be the duty of the Executive to make adequate arrangements for the enforcement of these Regulations, and accordingly a reference in the provisions applied for the purposes of such enforcement by sub-paragraph (b) below to an "enforcing authority" shall be construed as a reference to the Executive;

    (b) sections 19 to 28**(a)**, 33 to 35**(b)**, 38, 39, 41 and 42 of the 1974 Act shall apply for the purposes of providing for the enforcement of these Regulations and in respect of proceedings for contravention thereof as if–

        (i) references to relevant statutory provisions were references to those sections as applied by this paragraph and to these Regulations;

        (ii) references to articles, substances, articles and substances, or plant, were references to relevant products;

        (iii) references to the field of responsibility of an enforcing authority, however expressed, were omitted;

        (iv) in section 20, subsection (3) were omitted;

        (v) in section 23, subsections (3), (4) and (6) were omitted;

        (vi) in section 33–

            (aa) in subsection (1) the whole of paragraphs (a) to (d) were omitted;

            (bb) subsection (1A) were omitted;

            (cc) in subsection (2), the reference to paragraph (d) of subsection (1) were omitted;

            (dd) subsection (2A) were omitted;

            (ee) for subsection (3) there were substituted the following:–

                "(3) A person guilty of an offence under any paragraph of subsection (1) above not mentioned in subsection (2) above or of an offence under subsection (1)(e) above not falling within that subsection shall be liable–

                    (a) on summary conviction, to a fine not exceeding level 5 on the standard scale; or

                    (b) on conviction on indictment–

                        (i) in the case of an offence under subsection (1)(g), (j) or (o), to imprisonment for a term not exceeding two years, or a fine, or both; or

                        (ii) in all other cases, to a fine. "; and

            (ff) subsection (4) were omitted;

        (vii) in section 34–

            (aa) paragraphs (a) and (b) were omitted from subsection (1); and

            (bb) in subsection (3) for "six months" there were substituted "twelve months"; and

        (viii) in section 42, subsections (4) and (5) were omitted; and

    (c) sections 36(1) and (2) and 37 shall apply in relation to offences under section 33 as applied by sub-paragraph (b) above.

(a) In section 22, subsections (1) and (2) were amended and subsection (4) was added by paragraph 2, of Schedule 3 to, and section 36 of, the Consumer Protection Act 1987 (c.43). There is a modification of the application of section 24 not relevant to these Regulations. Sections 25A and 27A were inserted by paragraphs 3 and 4 respectively, and section 28(1)(a) was amended by paragraph 5, of Schedule 3 to, and section 36 of, 1987 c.43; section 27 was amended by the repeal of subsection (2)(b) and the word "or" immediately preceding it by section 29(3) and (4) of, and paragraph 10(1) and (2) of Schedule 6 and Schedule 7 to, the Employment Act 1989 (c.38), and in subsection (3) by section 33(1) of, and paragraph 7(a) of Part II of Schedule 3 to, the Employment Act 1988 (c.19) and section 29(3) of, and paragraph 10(3) of Schedule 6 to, 1989 c.38; and in section 28, subsections (3)(c) and (5)(b) were amended by section 190 of, and paragraph 46 of Schedule 25 to, the Water Act 1989 (c.15), a new subsection (6) was substituted by section 84 of, and paragraph 52 of Part II of Schedule 14 to, the Local Government Act 1985 (c.51), and new subsections (9) and (10) were added by section 116 of, and paragraph 9 of Schedule 15 to, the Employment Protection Act 1975 (c.71) and section 21 of, and paragraph 13 of Schedule 6 to, the Norfolk and Suffolk Broads Act 1988 (c.4) respectively.

(b) Section 33 was amended in subsection (1) in paragraph (h) by section 36 of, and paragraph 6 of Schedule 3 to, 1987 c.43, and in paragraph (m) by section 30 of, and Part I of the Schedule to, the Forgery and Counterfeiting Act 1981 (c.45); in subsection (2) as it applies to England and Wales by section 46 of the Criminal Justice Act 1982 (c.48); subsection (5) was repealed by section 4(5) of the Offshore Safety Act 1992 (c.15); and subsection (6) was repealed by section 30 of, and Part I of the Schedule to, 1981 c.45. There are other amendments to section 33, and there is an amendment to section 34, not relevant to these Regulations.

Forfeiture: England and Wales

2.—(1) An enforcement authority in England and Wales may apply under this paragraph for an order for the forfeiture of any relevant product on the grounds that there has been a contravention in relation thereto of regulation 6, 7 or 8.

(2) An application under this paragraph may be made–

    (a) where proceedings have been brought in a magistrates' court in respect of an offence in relation to some or all of the relevant products under regulation 16 to that court; and

    (b) where no application for the forfeiture of the relevant product has been made under sub-paragraph (a) above, by way of complaint to a magistrates' court.

(3) On an application under this paragraph the court shall make an order for the forfeiture of the relevant products only if it is satisfied that there has been a contravention in relation thereto of regulation 6, 7 or 8.

(4) For the avoidance of doubt it is hereby declared that a court may infer for the purposes of this paragraph that there has been a contravention in relation to any relevant products of regulation 6, 7 or 8 if it is satisfied that that regulation has been contravened in relation to a relevant product which is representative of that relevant product (whether by reason of being of the same design or part of the same consignment or batch or otherwise).

(5) Any person aggrieved by an order made under this paragraph by a magistrates' court, or by a decision of such court not to make such an order, may appeal against that order or decision to the Crown Court and an order so made may contain such provision as appears to the court to be appropriate for delaying the coming into force of an order pending the making and determination of any appeal (including any application under section 111 of the Magistrates' Courts Act 1980(**a**)).

(6) Subject to sub-paragraph (7) below, where any relevant product is forfeited under this paragraph it shall be destroyed in accordance with such directions as the court may give.

(7) On making an order under this paragraph a magistrates' court may, if it considers it appropriate to do so, direct that the relevant product to which the order relates shall (instead of being destroyed) be released, to such person as the court may specify, on condition that that person–

    (a) does not supply the relevant product to any person otherwise than–

        (i) to a person who carries on a business of buying relevant products of the same description as the first mentioned product and repairing or reconditioning it; or

        (ii) as scrap (that is to say, for the value of materials included in the relevant product rather than for the value of the relevant product itself); and

    (b) complies with any order to pay costs or expenses which has been made against that person in the proceedings for the order for forfeiture.

Forfeiture: Scotland

3.—(1) In Scotland a sheriff may make an order for forfeiture of any relevant product in relation to which there has been a contravention of any provision of regulation 6, 7 or 8–

    (a) on an application by the procurator-fiscal made in the manner specified in section 310 of the Criminal Procedure (Scotland) Act 1975(**b**); or

    (b) where a person is convicted of any offence in respect of any such contravention, in addition to any other penalty which the sheriff may impose.

(2) The procurator-fiscal making an application under sub-paragraph (1)(a) above shall serve on any person appearing to him to be the owner of, or otherwise to have an interest in, the relevant product to which the application relates a copy of the application, together with a notice giving him the opportunity to appear at the hearing of the application to show cause why the relevant product should not be forfeited.

(3) Service under sub-paragraph (2) above shall be carried out, and such service may be proved, in the manner specified for citation of an accused in summary proceedings under the Criminal Procedure (Scotland) Act 1975.

(**a**) 1980 c.43.
(**b**) 1975 c.21. Section 310 was amended by paragraph 53 of Schedule 7, and Schedule 8, to the Criminal Justice (Scotland) Act 1980 (c.62); there are extensions of section 310 not relevant to these Regulations.

(4) Any person upon whom a notice is served under sub-paragraph (2) above and any other person claiming to be the owner of, or otherwise to have an interest in, the relevant product to which an application under this paragraph relates shall be entitled to appear at the hearing of the application to show cause why the relevant product as the case may be should not be forfeited.

(5) The sheriff shall not make an order following an application under sub-paragraph (1)(a) above–

    (a) if any person on whom notice is served under sub-paragraph (2) above does not appear, unless service of the notice on that person is proved; or

    (b) if no notice under sub-paragraph (2) above has been served, unless the court is satisfied that in the circumstances it was reasonable not to serve notice on any person.

(6) The sheriff shall make an order under this paragraph only if he is satisfied that there has been a contravention in relation to the relevant product of regulation 6, 7 or 8.

(7) For the avoidance of doubt it is declared that the sheriff may infer for the purposes of this paragraph that there has been a contravention in relation to any relevant product of regulation 6, 7 or 8 if he is satisfied that that regulation has been contravened in relation to a relevant product which is representative of that relevant product (whether by reason of being of the same design or part of the same consignment or batch or otherwise).

(8) Where an order for the forfeiture of any relevant product is made following an application by the procurator-fiscal under sub-paragraph (1)(a) above, any person who appeared, or was entitled to appear, to show cause why it should not be forfeited may, within twenty-one days of the making of the order, appeal to the High Court by Bill of Suspension on the ground of an alleged miscarriage of justice; and section 452(4)(a) to (e) of the Criminal Procedure (Scotland) Act 1975(a) shall apply to an appeal under this sub-paragraph as it applies to a stated case under Part II of that Act.

(9) An order following an application under sub-paragraph (1)(a) above shall not take effect–

    (a) until the end of the period of twenty-one days beginning with the day after the day on which the order is made; or

    (b) if an appeal is made under sub-paragraph (8) above within that period, until the appeal is determined or abandoned.

(10) An order under sub-paragraph (1)(b) shall not take effect–

    (a) until the end of the period within which an appeal against the order could be brought under the Criminal Procedure (Scotland) Act 1975; or

    (b) if an appeal is made within that period, until the appeal is determined or abandoned.

(11) Subject to sub-paragraph (12) below, relevant products forfeited under this paragraph shall be destroyed in accordance with such directions as the sheriff may give.

(12) If he thinks fit, the sheriff may direct the relevant product to be released to such person as he may specify, on condition that that person does not supply it to any person otherwise than–

    (a) to a person who carries on a business of buying relevant products of the same description as the first-mentioned relevant product and repairing or reconditioning it; or

    (b) as scrap (that is to say, for the value of materials included in the relevant product rather than for the value of the relevant product itself).

Duty of enforcement authority to inform Secretary of State of action taken

4. The enforcement authority shall, where action has been taken by it to prohibit or restrict the placing on the market, the supply or putting into service (whether under these Regulations or otherwise) of any relevant product which bears the CE marking forthwith inform the Secretary of State of the action taken, and the reasons for it, with a view to this information being passed by him to the Commission.

Savings

5. Nothing in these Regulations shall be construed as preventing the taking of any action in respect of any relevant product under the provisions of the 1974 Act.

6. Nothing in these Regulations shall authorise the enforcement authority to bring proceedings in Scotland for an offence.

(a) A new section 452 was substituted by paragraph 11 of Schedule 3 to 1980 c.62.

Interpretation

7. In this Schedule–

"the 1974 Act" means the Health and Safety at Work etc. Act 1974(a);

"the Executive" means the Health and Safety Executive established under section 10 of the 1974 Act; and

"relevant product" means an item of equipment, a protective system, a device or component, as the case may be, to which these Regulations apply.

(a) 1974 c.37.

SCHEDULE 15

Regulation 20(2)(c)

DISAPPLICATION OF SPECIFIC MINING REGULATIONS(a)

PART I

(Made under the Mines and Quarries Act 1954(b))

(1) Title	(2) Reference	(3) Extent of disapplication
The Woodside Nos 2 and 3 Mine (Diesel Vehicles) Special Regulations 1960	S.I. 1960/1291	Regulations 4 to 6
The Grimethorpe Mine (Diesel Vehicles) Special Regulations 1961	S.I. 1961/2444	Regulations 4 to 6
The Lynemouth Mine (Diesel Vehicles and Storage Battery Vehicles) Special Regulations 1961	S.I. 1961/2445	Regulations 4 to 7
The Calverton Mine (Diesel Vehicles) Special Regulations 1962	S.I. 1962/931	Regulations 4 to 6
The Brightling Mine (Diesel Vehicles) Special Regulations 1962	S.I. 1962/1094	Regulations 5 to 7
The Easington Mine (Diesel Vehicles) Special Regulations 1962	S.I. 1962/1676	Regulations 4 to 6
The Rufford Mine (Diesel Vehicles) Special Regulations 1962	S.I. 1962/2059	Regulations 4 to 6
The Trelewis Drift Mine (Diesel Vehicles) Special Regulations 1962	S.I. 1962/2114	Regulations 4 to 6
The Wharncliffe Woodmoor 4 and 5 Mine (Diesel Vehicles) Special Regulations 1962	S.I. 1962/2193	Regulations 4 to 6
The Seaham Mine (Diesel Vehicles) Special Regulations 1962	S.I. 1962/2512	Regulations 4 to 6
The Dawdon Mine (Diesel Vehicles) Special Regulations 1963	S.I. 1963/118	Regulations 4 to 6
The Thoresby Mine (Diesel Vehicles) Special Regulations 1963	S.I. 1963/825	Regulations 4 to 6
The Westoe Mine (Diesel Vehicles) Special Regulations 1963	S.I. 1963/1096	Regulations 4 to 6
The Silverwood Mine (Diesel Vehicles) Special Regulations 1963	S.I. 1963/1618	Regulations 4 to 6
The Prince of Wales Mine (Diesel Vehicles) Special Regulations 1964	S.I. 1964/539	Regulations 4 to 6
The Cwmgwili Mine (Diesel Vehicles) Special Regulations 1964	S.I. 1964/1225	Regulations 4 to 6
The Wearmouth Mine (Diesel Vehicles) Special Regulations 1964	S.I. 1964/1476	Regulations 4 to 6
The Bevercotes Mine (Diesel Vehicles) Special Regulations 1965	S.I. 1965/1194	Regulations 4 to 6
The Ellington Mine (Diesel Vehicles and Storage Battery Vehicles) Special Regulations 1967	S.I. 1967/956	Regulations 4 to 7
The Prince of Wales Mine (Captive Rail Diesel Locomotives) Special Regulations 1969	S.I. 1969/1377	Regulation 4
The Boulby Mine (Storage Battery Locomotives) Special Regulations 1972	S.I. 1972/472	Regulations 4 to 8
The Elsecar Main Mine (Diesel Vehicles) Special Regulations 1974	S.I. 1974/710	Regulations 4 to 6

(a) The regulations specified in the third column to this Schedule are also revoked, with effect from 1st July 2003, by regulation 1(2) of and Schedule 1 to these Regulations.

(b) 1954 c.70.

PART II

(Made under the Mines and Quarries Act 1954 and the Health and Safety at Work etc. Act 1974(**a**))

(1) Title	(2) Reference	(3) Extent of disapplication
The Rixey Park Mine (Storage Battery Locomotives) Special Regulations 1974	S.I. 1974/1866	Regulations 4 to 8

PART III

(Made under the Health and Safety at Work etc. Act 1974)

(1) Title	(2) Reference	(3) Extent of disapplication
The Markham Mine (Diesel Vehicles) Regulations 1976	S.I. 1976/1734	Regulations 4 to 6
The Bentinck Mine (Diesel Engined Stone Dusting Machine) Regulations 1976	S.I. 1976/2046	Regulations 5 to 7
The Thoresby Mine (Cable Reel Load-Haul-Dump Vehicles) Regulations 1978	S.I. 1978/119	Regulations 4 to 9
The Trelewis Drift Mine (Diesel Vehicles) Regulations 1978	S.I. 1978/1376	Regulations 4 to 6
The Boulby Mine (Diesel Vehicles) Regulations 1979	S.I. 1979/1532	Regulations 5 to 8
The Harworth Mine (Cable Reel Load-Haul-Dump Vehicles) Regulations 1980	S.I. 1980/1474	Regulations 4 to 9
The Point of Ayr Mine (Diesel Vehicles) Regulations 1980	S.I. 1980/1705	Regulations 5 and 6

(**a**) 1974 c.37.

EXPLANATORY NOTE

(This note is not part of the Regulations)

These Regulations implement, in Great Britain, the European Parliament and the Council Directive 94/9/EC on the approximation of the laws of the Member States concerning equipment and protective systems intended for use in potentially explosive atmospheres (the ATEX Directive). Subject to certain limited exceptions, references in the Regulations to the Community or a member State include respectively a reference to the EEA or an EEA State, which are defined in regulation 2(1)(b).

Part II provides for the application of the Regulations. They apply to equipment and protective systems intended for use in potentially explosive atmospheres, devices and components (relevant definitions are contained in regulation 3), other than those excluded by regulation 4 or 5. The Regulations do not apply to the equipment, protective systems or devices specified in Schedule 5 or components for them (regulation 4). Regulation 5 sets out transitional arrangements whereby the Regulations do not apply to equipment or a protective system placed on the market in the Community/EEA on or before 30th June 2003 which complies with the health and safety provisions in respect of it which were in force in Great Britain on 23rd March 1994.

Part III sets out the general requirements of the Regulations. Regulation 6 imposes a duty on a "responsible person" (as defined in regulation 2(2)) who places on the market equipment, a protective system or device to ensure that it satisfies the relevant essential health and safety requirements and that the appropriate conformity assessment procedure has been carried out. In addition, the CE marking must be affixed to it by the manufacturer or his authorized representative in the Community/EEA in accordance with Schedule 2. The relevant product must also be safe. Any person, other than the responsible person, who supplies equipment, a protective system or device must ensure that it is safe (regulation 7) although the regulation does not apply to such a product if it has been placed on the Community/EEA market before 1st March 1996 or has previously been put into service in the Community/EEA.

Regulation 8 requires a responsible person who places a component on the market to ensure that the appropriate conformity assessment procedure has been carried out and that it is accompanied by a certificate, issued by the manufacturer or his authorized representative established in the Community/EEA, which incorporates the specified declaration and statement.

For the purposes of regulations 6 and 8, a product is not to be regarded as being placed on the market where it is for export outside the Community/EEA or is exhibited at a trade fair or exhibition (regulation 9).

The essential health and safety requirements (Annex II of the ATEX Directive) are set out in Schedule 3. Regulation 2 defines "relevant essential health and safety requirements".

The conformity assessment procedures are set out in various "Modules" in Annexes III to IX of the ATEX Directive (which are set out in Schedules 6 to 12). The appropriate conformity assessment procedure is determined in accordance with regulation 10, on the basis of the equipment-group (as defined in regulation 2(2)) and the equipment category (as determined by the criteria in Schedule 4) of the product.

Regulation 12 provides for the appointment of notified bodies in Great Britain and specifies their functions. Regulation 13 provides for these bodies to charge fees.

Regulation 14 sets out the conditions for relevant products being taken to comply with the provisions of the ATEX Directive.

Part IV and Schedule 14 make provision for the enforcement of the Regulations by the Health & Safety Executive. In Scotland, proceedings are brought by the procurator-fiscal or Lord Advocate. Regulations 16 and 17 provide for the offences and penalties for breach of the Regulations. There are also provisions relating to the defence of due diligence (regulation 18) and liability of persons other than the principal offender (regulation 19).

Certain amendments and disapplications of the law in Great Britain are made by regulations 20 and 21. These include an amendment to The Electricity at Work Regulations 1989 which is required in order to incorporate a reference to Commission Directives

91/269/EEC and 94/44/EC which adapted to technical progress Council Directive 82/130/EEC concerning electrical equipment for use in potentially explosive atmospheres in mines susceptible to firedamp (regulation 20(1)). Regulation 20(2) provides for the disapplication of specified Mine regulations, which are set out in that paragraph and in Schedule 15: the majority of those specified regulations are revoked with effect from 1st July 2003 (regulation 1(2) and Schedule 1). Regulation 21 makes a consequential amendment to the Provision and Use of Work Equipment Regulations 1992 and provides that these Regulations are to take effect for the purposes of the enforcement of regulation 10 of the 1992 Regulations as if the amendment had been made under section 15 of the Health and Safety at Work etc. Act 1974.

Schedule 1 also sets out other Regulations which are revoked with effect from 1st July 2003. Schedule 2 also includes provisions relating to inscriptions other than the CE marking. Schedule 13 specifies the content of the EC declaration of conformity.

A Compliance Cost Assessment in respect of these Regulations is available and a copy can be obtained from the Department of Trade and Industry, Standards Policy Unit, 3rd floor, 151 Buckingham Palace Road, London SW1W 9SS.

1996 No. 192

HEALTH AND SAFETY

The Equipment and Protective Systems Intended for Use in
Potentially Explosive Atmospheres Regulations 1996